Ubiquitous Science

Let's explore

A K Bala Muruhan

This book Ubiquitous Science is authored and published by A K Balamuruhan

Copyright © 2019 by A K Balamuruhan All rights reserved

It is not legal to reproduce, duplicate or transmit any part of this document in either electronic means or printed format.

I dedicate this book to my teachers, my family members, my friends and the readers of the book.

Contents

Preface	ix

1 Philosophy of creation   1
   1.1 The philosophy 1
       1.1.1 Simple arithmetic 1
       1.1.2 New arithmetic 3
   1.2 Evidence from quantum physics 4
       1.2.1 The source 5
   1.3 Creation of physical entities 6
       1.3.1 Creation of space 6
       1.3.2 The Force 7
       1.3.3 Creation of particles - a speculation 8
   1.4 What is quantum spin? 9
       1.4.1 The two spin states 10
       1.4.2 Mathematical representation of spin flip . . . 12

2 Analysing the philosophy   15
   2.1 Bipolarity, the Super symmetry 15
   2.2 Laws created and laws of creation 16
   2.3 Conservation? 16
       2.3.1 Dark energy 17
   2.4 Spin 17
       2.4.1 Quantum spin and the classical dynamics . . . 17
       2.4.2 Calculating the angular momentum 19
   2.5 Gravity and its effects 21
       2.5.1 Gravity being explained 23
       2.5.2 Interpreting curvature of spacetime 26
       2.5.3 Corroborating with our philosophy 30
       2.5.4 Contraction of space by matter 31

		2.5.5 Gravitational and inertial masses	32
	2.6	What is more?	33

3 The Science Beyond — 35
3.1 The purpose of this chapter 35
3.1.1 Limits of physical sciences 36
3.2 The scripture 36
3.3 Saiva Sithandham 37
3.3.1 God's creations from Maya 40
3.3.2 Spiritual path 43
3.3.3 Can it be a probabilistic evolution? 46
3.3.4 Withdrawal of creations? 47

4 Overview — 49

List of Figures

1.1	Basic arithmetic in length	2
1.2	Variation of the intensities of the squares of the two parts of the wave function.	10
2.1	The orbital speeds (rotation curve) of planets in our solar system	21
2.2	Rotation curve of galaxy M33	22
2.3	The rotation curves of a typical galaxy	22
2.4	World-line examples	27
2.5	An object accelerating towards another	27
2.6	Acceleration between two objects as seen from reference frame of B	28
2.7	Constant position lines of a 'flat' space	29
2.8	The constant position lines curved by the gravity of object A	30
2.9	Refraction of light by densification of space quanta	31
2.10	Refractive index	32
3.1	Lord Dhakshinamoorthy	38
3.2	Spiritual stages	43
3.3	Grasshopper with neem-leaf structure	46
3.4	An insect looking like a dry mango leaf	47

Preface

There are many unanswered questions in Science. By science, I mean here physical science. For example, the singularity before the origin of the universe is an unsolved problem; Dark matter and dark energy appear to remain mysteries; In the microscopic world, it is not clear what the quantum mechanical spin means. Let's see whether we can unravel the mysteries behind some of such issues. Let's approach with an open mind and evaluate some new concepts that will be discussed in this book, which I conceived during the course of past twenty or more years. The concepts are built upon introductory concepts in Physics. Hence, this book is intended for readers with preliminary knowledge in physics and also readers with expertise in different fields of physics, who can evaluate the concepts.

Chapter 1 introduces the concepts mentioned above. The fields of these concepts range from atomic level physics to cosmology. The discussion on *Time* becomes abstract, which is apparently inevitable. Some concepts introduced are deep rooted in me; some are speculative.

Chapter 2 discusses a few consequences of the philosophy introduced in chapter 1. In this process, this chapter also introduces a few new concepts, which again need evaluation by experts in the corresponding fields of physics.

Chapter 3 introduces and discusses the Science beyond physical sciences. In my opinion, the domain of this science includes physical sciences also to some extent although its emphasis is not on this. Science is everywhere, ubiquitous. I hope I have provided proper interface between this chapter and the previous two chapters.

Chapter 1

Philosophy of creation

The conventional idea about the creation of the universe is that the whole universe with all its physical contents was concentrated at a single point before the creation started. This could be a myth since it involves infinite mass energy at that point, which is not physically sensible. This chapter introduces some new concepts of creation of the physical universe. One of the advantages with these new concepts is that the oddness of singularity discussed above is removed. Further advantages will be visible as we will go along. For example, a solution to dark matter appears as a consequence of these concepts. Care is taken to present the concepts as clear as possible. However, contemplation may be required to comprehend the concepts. Please let your mind expand into new horizons.

A new philosophy of creation was budding up as a result of my attempt to interpret the quantum mechanical wave functions. The philosophy is presented first and then the evidences from the wave functions are discussed.

1.1 The philosophy

1.1.1 Simple arithmetic

We know addition and subtraction as two basic arithmetic. What do they mean?

- **In space:** Let us say three persons stand as shown in figure 1.1. We may say person B is at a distance of 6 steps from person A and person C is at a distance of -2 steps from person A meaning

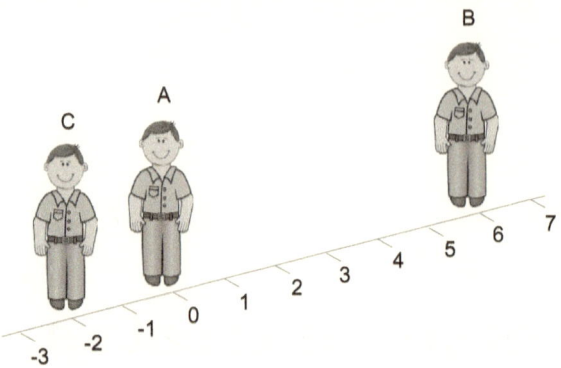

Figure 1.1: Basic arithmetic in length

that he is at a distance of 2 steps from person A in the direction opposite to that in which person B is. If person B takes steps towards person A, with each step the distance between them decreases by one unit. When he reaches person A, the distance between them is zero steps. If he continues to walk past person A, say by one step, his distance from person A becomes -1 step. With every step he takes in the same direction, his distance is told to be more and more negative from A.

- **In time** Let 6'O clock in the morning be the time you wake up; You set off to office at 8AM, return home at 6PM. The time gap of your setting off to office from waking up is 2 hours. The time gap of your waking up from returning home is -12 hours.

- **matter** Suppose person A has two apples. Let B have three apples. Their total possession is five apples. The concept of negative quantity of matter can be brought in if phenomena like loan and owing are invoked. Let A borrow one apple from B. Now they have 3 and 2 apples respectively. After they consume the apples, their possession is zero apples in hand each. Considering the one apple A owes to B, the possession of A is told to be -1 apple and the possession of B is told to be one apple; this is notional. This will come in reality after A repays the apple. Suppose A and B earn three apples each and A repays the apple. Now A and B will have two and four apples respectively.

1.1. THE PHILOSOPHY

Why do we discuss this arithmetic now? It is to take note of what positive and negative quantities mean in space and time (and matter too) in the conventional sense. This arithmetic involve quantities and operations relative to some reference that is free to be varied. Now, let us move on to the more absolute arithmetic.

1.1.2 New arithmetic

- **Matter** Let there be two apples in a box. You take one from the box. Now you have one in the hand and one in the box, the total being two. You take one more apple from the box. You have two in the hand and zero in the box, the total again being two. Now, you take one more apple from the box. One more apple? How is it possible? May be, it is not possible. If you somehow, are able to take one more apple from the box, what is the mathematical result? You have three apples in the hand and -1 apple in the box, the total number being -1+3=2. If there had been zero apple originally in the box, you would now have one apple in the hand and -1 apple in the box. -1 apple in the box! What does this mean? This may not have a physical meaning. However, if it can happen really, the one in the box should also be an apple; because only then, can it nullify the one in the hand to produce zero. Let us not further ponder on the physical characteristics of these apples.

- **Space and Time** We imagined creation of matter with reference to absence of matter (zero apple in the box). Absence of matter is easy to comprehend. Absence of space is not that easy a concept to comprehend. We postpone it until we discuss creation of physical entities in the next section. Discussion on time follows thereafter.

Although we do not give physical meaning to the pair of apples, let us make here a huge step by assuming that this is the basis of creation of all physical entities in the universe. Does this mean that every macroscopic object like an apple is created with its counter-object, a counter-apple? Obviously, No, since we do not find any such pairs of counter-objects in nature. Then, can we imagine that every fundamental particle like an electron was created with its counter-particle? Again, No, since we do not find such pairs of counter particles. The electron-positron pair cannot be considered as one such pair, since an electron and a positron annihilate each other into an equivalent amount of energy, not into 'nothing'. In our arithmetic with apples, the pair of apples combine to form zero, not any amount of energy; neither did we pump in that much energy to produce the pair of apples.

Then, where can we find such pairs? Let's assume that every fundamental particle is created with two such parts within itself. The particle is constituted by two mutually opposite parts that can annihilate each other into 'nothing' and were created from that 'nothing'. Where do we find the evidence for this?

1.2 Evidence from quantum physics

Quantum physics is the science of micro world. In quantum physics, every fundamental entity like an electron or an atom is described by its wave function. A wave function is complex natured; it has a real part and an imaginary part[1].

We interpret that the real part represents one part of the entity and the imaginary part represents the other part of the entity. To be more precise, the square of the real part describes the spatial and temporal spread of one part and the square of the imaginary part describes the spatial and temporal spread of the other part.

Square of the real part is positive and square of the imaginary part is negative. Now, recall our imagination with apples. We had one in hand and -1 in the box. The opposite signs mentioned here represent those opposite signs. The physical existences of the two parts are mutually opposite.

This serves the conservation. Before creation, the entity did not exist; after creation, the entity exists. The mutual oppositeness in the physical existences of the two parts conserves the physical-non-existence before creation[2].

---

[1]$x + iy$ is a complex number, where x and y are real numbers and i is defined as $\sqrt{-1}$ or $i^2 = -1$. Square of any (positive or negative) real number is a positive number. Hence, square root of a negative number cannot be a real number. Hence, it is called an imaginary number. $\sqrt{-1}$ is known as imaginary unit. Hence, in $x + iy$, x is known as its real part and iy is known as its imaginary part. $x - iy$ is known as its complex conjugate. Why such a thing called as complex conjugate?? It will be discussed later.

[2]Conventional science does not have an answer to the question, "Where from does the physical content of the universe come?". It tells that it existed forever. How can it? An origin is a legitimate requirement. If the physical content of the universe is constant, what is its

1.2.1 The source

Although we have introduced a fundamental conservation, there is still something lacking. How can something be created from nothing? Does it not breach conservation? Yes, it does. Something is required to give birth to these physical entities. What is that something? We do not expect it to be nothing. Neither do we expect it to be something physical because that is our basic supposition. It is not nothing and it is not physical. Hence, it has to be something subtle. We call it the 'subtle source'. In the next section, let us see how things are created from this subtle source.

Before moving on to the creation of physical entities, we have to discuss one more important characteristic of the two parts. Recall that we expected the -1 apple in the box also to be an apple, because only then will it be able to nullify the apple in the hand to form zero. Likewise, we expect the two parts of the fundamental entity, an electron for example, are equivalent. Neither of them is more physical than the other; neither of them is more abstract than the other. Please do not be misled by the negative sign with one of the two parts. This may induce to think of that part to be more abstract as the -1 apple in the box would normally be imagined as an abstract object. With a macroscopic object, it may be so; however, with the fundamental entities, it need not be so. Both can be equally physical; however, remember that the nature of their physical existences are mutually opposite. This is a new concept. Hence, it may require time to contemplate. Based on the above discussions we propose the following hypotheses.

Hypothesis 1: *Every fundamental physical entity is created from a subtle source, with two parts, which are inherently opposite to each other in their nature of existence and which are equally physical. The entity was present subtly in the source; creation is the act of transformation from that subtle existence to physical existence.*

Hypothesis 2: *The absolute magnitudes of the squares of the real and imaginary parts of the quantum mechanical wave function of a physical entity describe the spatial distribution and temporal behaviour of the two parts of the entity, the opposite signs of the squares indicating the mutual oppositeness in the fundamental nature of existence of the two parts.*

---
quantity? What is that magic number? Is there any uniqueness about that number?

Hypothesis 3: *The potential of the subtle source is infinite to create any quantities of physical entities.*

Every physical entity has physical attributes like mass. The mutual oppositeness of the two parts means that these attributes of the two parts are mutually opposite; however, they are equally physical [3]. Then why do we not perceive these two types of parts in the physical world? It is because everything in the universe consists of these two types; the observer, the observed and the tool for observation, all these have these two types of parts inherently; and, we have not yet known the laws that govern the interaction between the like-parts and opposite parts of the interacting entities; if we know and if they are different, we can devise experiments to sense the difference between these two types of interactions and thereby, verify the existence of the two parts of physical entities. Let us accept the hypothesis until then and proceed.

Since we take the two parts of the wave function of a particle to represent the two equivalent parts of the particle, the nomenclature of the two parts to call one of them as 'real part' and the other as 'imaginary part' is misleading. It's preferable to call the two parts of the wave function by names such as part 1 and part 2. Mathematically, the two parts of the wave function are interchangeable in quantum mechanics. We will see in section 1.4.2 what it means in our description by the exchange of the real and imaginary parts of a wave function.

1.3 Creation of physical entities

Now, let's start discussing the creation of physical entities one by one. The space is the most fundamental creation. It is the placeholder for all the other physical entities. Hence, we start with space.

1.3.1 Creation of space

Space can be assumed as emanating from the subtle source. Space provides the properties, location and dimensions to other physical entities. The subtle source is not bound by any of these. Why?

---

[3] We repeatedly mention that the two parts are equally physical to emphasize that either of them is not more physical or more abstract than the other.

1.3. CREATION OF PHYSICAL ENTITIES

Because it is the provider of space itself. It is beyond space. This means that it is everywhere in space. This is how further space can be created from subtle source anywhere in space. We can also interpret it that every quantum of space is associated with a quantum of subtle source proportional to its volume.

Conventional cosmology assumes that space started expanding from the primeval point. It assumes a history of expansion based on different cosmological observations and existing theories of physics. This may be correct or need modification. Whatever the correct history of expansion, our philosophy would interpret it that space was created (not expanded) in such a way.

Space also being a created entity, it should also have two mutually opposite parts, according to hypothesis 1! The assimilation of this concept may be difficult. We will get more clarity on it when we discuss spin in section 1.4.1.

1.3.2 The Force

Now, let's introduce an important concept. We propose that every quantum of subtle source exerts a force of attraction with every other quantum of source. This will be viewed as every quantum of space exerting a force of attraction with every other quantum of space. Thus, the created space shrinks in response to this force. The more a quantum of space is interior in a created volume of space, the more is the pressure it experiences by the surrounding space and the more it shrinks. This is like the pressure inside the sun increasing towards its centre. Thus, the density of space and subtle source in a region of space is proportional to the amount of space surrounding it. We will see later that it is also proportional to the amount of matter surrounding it.

We propose that this force is the force of gravity. Gravity with space? We will get more clarity on this in due course.

Postulate 1: *Gravity is the force of attraction exerted by every quantum of subtle source on every other quantum of subtle source.*

How does this definition of gravity fit the already known facts about gravity that it acts between matter? This will be discussed in the next section.

1.3.3 Creation of particles - a speculation

We introduce the postulate on creation of particles.

Postulate 2: *When the shrinking of a volume of space caused by the gravitational pressure exerted by its surrounding exceeds a limit, a particle starts emanating from its centre by the conversion of space there into the particle and a quantum of space gets converted into a complete particle. This causes a flow of space towards this point.*

A particle, as we introduced in section 1.2, is an extended entity described by its wave function. Thus, particles that are created from space, then occupy space. Such conversion of space into particles will continue as long as the pressure in a region exceeds the limit.

Gravitational force with matter: The quantum of subtle source that was associated with the quantum of space from which a particle was created is now associated with that particle. The quantum of subtle source continues to exert its gravitational force of attraction. Thus, whatever force of attraction was exhibited by that quantum of space is now exhibited by the particle created from it. This force of attraction acting between macroscopic objects constituted by such particles was identified as force of gravity by Newton. This aspect will be further discussed in section 2.5.

The particles are created with the three other fundamental forces of nature and experience the corresponding potential with each other, which may result in kinetic energy too. We do not know whether the particles are created with an initial kinetic energy.

Hypothesis 4: *Gravity holds a special place among the four fundamental forces of nature, since gravity is inherent to and acts between the parts of the subtle source while the other three forces are created from the subtle source to act between the particles.*

The interpretation of a wavefunction as describing the extended existence of a particle contrasts the conventional interpretation that the amplitude of a wave function in a region of space represents the probability of finding the whole of that particle in that region. According to our interpretation, 'finding a particle in a region of space' implies that the transformation regarded as 'finding' (which could be a process like scattering or sharing of energy) is initiated by the overlap in that region of space between the extended existence of the 'probe that finds' and the extended existence of the 'particle being found'. Thus, according to this picture, particles do not have wave-particle duality; They have only wave nature; The exhibit of 'particle nature' is only a wave process.

Consequently, the entities like electromagnetic radiation also do not have particle nature. They only have wave nature. The interaction between a quantum of electromagnetic radiation and a particle like an electron is also a wave process as interpreted above.

1.4 What is quantum spin?

There is a concept known as the spin of a particle in quantum mechanics. It is considered so far, as an abstract concept inherent to a particle, without any physical visualization possible. Let's provide it a clear visualization. We know that the wave function is time dependant. With time, the intensities of the squares of the two parts of the wave function increase and decrease. The increasing and decreasing of these two quantities complement each other as shown in figure 1.2. When one increases, the other decreases and vice versa, conserving the (absolute) sum of their intensities along time as seen in the figure; this conservation means the conservation of the physical existence of the physical entity that the wave function represents. The absolute sum is achieved by multiplying the wave function with its complex conjugate. This is the purpose of complex conjugation in quantum mechanics. It signifies the equivalence between the two parts, as discussed earlier. We hypothesize the quantum spin as below.

Postulate 3: *The cyclic intensification and rarefaction of the two parts of a particle described by the squares of the two parts of its wave function is known as the quantum spin of that particle.*

This means that the two parts of a particle at any point of space come into existence and go into absence cyclically and complementarily and this process is known as spin.

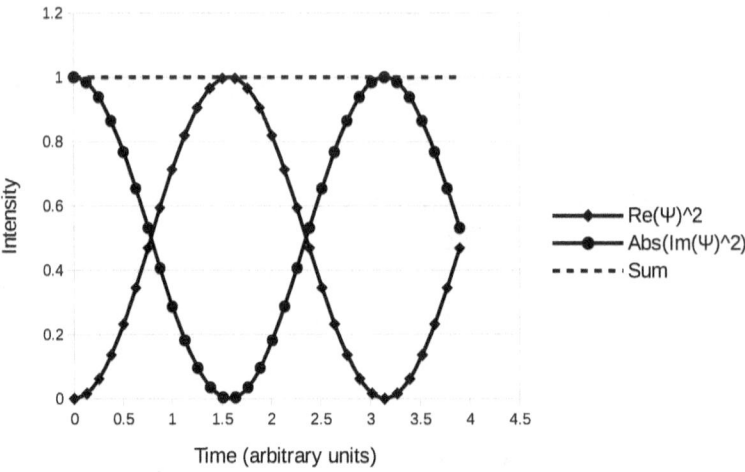

Figure 1.2: Variation of the intensities of the squares of the two parts of the wave function.

1.4.1 The two spin states

Quantum mechanically, a particle is said to be in either of two spin states, either in the spin up state or in the spin down state. We hypothesize on these two spins as below.

Hypothesis 5: *The intensification of part 1 of a particle coupled with the rarefaction of part 2 can be regarded as one spin state, say, the spin up state; then the transformation in the opposite direction, viz., rarefaction of part 1 coupled with the intensification of part 2 will be regarded as the spin down state.*

A notion on Time After an intensification and rarefaction of the two parts of a particle is completed, these two parts start retreating, that is, they start to rarefy and intensify respectively. Does this mean that the direction of the spin of the particle flips at this moment? The answer should be No, since observation tells us that the spin of a particle doesn't flip this way. Then how to comprehend this? For this, we need a special notion of time, which is very abstract. We hypothesize on time as below.

Hypothesis 6: *Since time is a created entity, it should also have two (mutually opposite) components, like all other physical creations have.*

How does this hypothesis help resolving the conflict of continuous spin flip of a particle, discussed above? Let's suppose that a particle is in up spin state meaning that its part 1 intensifies and its part 2 rarefies with time; Time also having two parts, let's say that this action 'happens along with part 1 of time'. Further, we say that the subsequent retreat of rarefaction and intensification happens with part 2 of time. This part of time being opposite to part 1, the reversal of direction happening with the retreat of the particle gets nullified, just as two negative signs negate each other and result in positive sign. This concept appears abstract, and in fact, vague. Still, it portrays some fact. Please get along with the levels of subtleties involved here.

Two spins at a place

Consider a moment at which the intensity of part 1 of a particle at a point in space is at its maximum; then, the intensity of its part 2 at that point is zero. Then, with time, part 1 will start to vanish while part 2 starts to appear, at that point. Recall that space also has two mutually opposite parts. Let us say that part 1 of the particle occupies part 1 of space and part 2 of the particle occupies part 2 of space. This implies that the occupancy of the two parts of space increases and decreases cyclically and complimentarily. More importantly, this implies that the remaining fractions of the two parts of space is available for occupation by the two parts of another particle. The intensities of the two parts of the second particle will complement the intensities of the two parts of the first particle and this means that the second particle is 'spinning' in opposite direction to that of the first particle. This is how two particles and only up to two particles can occupy the same space, if their spins are opposite! This is what is required by Pauli exclusion principle. We are also now clear why there are only two spin states possible for a particle.

Source of Time

Each physical particle is attached to a quantum of the subtle source, as seen in section 1.3. We know that there are innumerous instances of physical particles and hence, there are innumerous quanta of subtle source. However, all of them are driven by a single instance of time, which is common to all the particles. Hence, we hypothesize on the source of time as below.

Hypothesis 7: *Time has to be created from a source that is subtler than the subtle source of other physical entities.*

There is one instance of time driving innumerous instances of particles. This hints at the possibility of 'action at a distance'. In fact, the subtle source of the particles is also present everywhere. Consider a pair of 'entangled' particles. If the spin of a particle flips, the spin of the other particle has to flip instantaneously, according to quantum mechanics. Such an instantaneous response is known as action at a distance and could be possible through the single entity of time connecting both these particles through the subtle source of these particles. Wootters[4] also identifies the time factor in wave functions as responsible for the entanglement and 'action at a distance', in his work on real vector space quantum mechanics. That work considers both parts of the wave function to be real.

Time is uppermost in the hierarchy of creations. The next are the laws of nature. The other physical entities are still lower in hierarchy. Time may be considered as a subtle creation. Time activates the physical entities according to these laws of nature. If the response to an act has to be instantaneous (as with entangled particles' 'action at a distance'), time executes it instantaneously. At situations where laws of relativity have to be obeyed, time enacts the scenes accordingly.

1.4.2 Mathematical representation of spin flip

For the spin to flip, the two parts of the particle should transform in the opposite direction. This is achieved by multiplying the wavefunction by i or $-i$. If the wavefunction is $\psi = \psi_r e^{i\omega t}$, then $i\psi = \psi_r e^{i\left(\omega t + \frac{\pi}{2}\right)}$ represents the particle with spin flipped. The squares of the real and imaginary parts of this wave function are shifted by a phase angle of π in their time evolution, which represents the two parts transforming in the opposite direction, which, in turn, means flipping of spin as discussed in section 1.4.1.

---
[4] https://arxiv.org/abs/1210.4535

1.4. WHAT IS QUANTUM SPIN?

If two particles in a system flip their spins, then the wave funcion of the system has to be multiplied by $i^2 = -1$. This explains the requirement of Pauli exclusion principle that the total wave function is antisymmetric with respect to exchange of two identical particles.

We have introduced some new philosophies and discussed a few aspects of the them in this chapter. In the next chapter, let us analyze the philosophies a little more.

Chapter 2

Analysing the philosophy

What are the consequences of the philosophies discussed in chapter 1? Let's analyze a few of them a little and leave a lot for the experts in different fields of physics to ponder upon further.

2.1 Bipolarity, the Super symmetry

We found that every fundamental physical entity is created with two parts that are mutually opposite in their nature of physical existence. This bipolarity of existence seems to be the Supersymmetry of all other symmetries discussed in Science.

We see in an atom, the positive charge of the protons and the negative charge of the electrons to be symmetric about a neutral charge, possessed by neutrons. However, the bipolar picture implies that within an electron, there are two parts, both having a negative charge, which are however fundamentally opposite in their nature of existence. This means that the negative electric charge of these two parts are fundamentally opposite in their nature of existence. This idea could be strange for the time being. Please get used to it.

For the physical analysis of the created world, both the parts can be (or have to be) considered equivalently. This is achieved by multiplying the wave function with its complex conjugate, while evaluating any physical quantity.

2.2 Laws created and laws of creation

We saw that there was no space before the 'Big bang' and space is created from the subtle source. However, we did not propose any particular pattern of creation of space. We call the law that governs this creation, a law of creation, since it involves creation of entities of the universe. It is apparent that the laws of creation are not known to us.

The laws created are the laws that are created to govern how the created entities perform in the created universe. Examples are the physical laws such as laws of gravity, thermodynamics, relativity and so on.

2.3 Conservation?

According to conventional belief, the universe started with a process called 'Big Bang'; Before 'Big Bang', the whole universe was believed to be condensed at a point; hence, the mass-energy density at that point is believed to have been infinite; hence, this condition is seen as a singularity.

According to the philosophy of creation proposed in chapter 1, the universe was absent in its physical form before the creation of universe started. The creation of the universe started with the creation of space and then continued with the creation of particles and other aspects of physical universe, from the subtle source. This creation continued and still continues (which we will address in section, 2.3.1) and in this process, the physical content of the universe increases with time. There was no singularity before the start of the creation. The total physical content or mass-energy content of the universe was nil before the creation and it continued to increase with the creation. Thus the mass-energy content of the universe is not conserved. It has ever been increasing. In principle, a created physical entity can be taken back into its source, an event of which has not probably happened or has not been observed so far.

However, the mass-energy is conserved in processes that happen completely within the domain of created universe. For example, in processes involving mass to energy conversion like matter-antimatter annihilation or nuclear fission happening in nuclear reactors, the mass-energy is conserved. Mass energy is also conserved in the inverse processes involving energy to mass conversion like pair production where a gamma ray gets converted into an electron and a positron. Mass energy is also conserved in processes where one form of physical energy is converted into another form. In short, the laws of mass-energy conservation hold good in all processes where creation or annihilation of physical entities as we have introduced is not involved.

2.3.1 Dark energy

The expansion of the universe is caused by the creation of space. In fact, the expansion has been observed to be accelerating. Saul Perlmutter, Brian P. Schmidt and Adam G. Riess were awarded Nobel prize in the year 2011 for this discovery of accelerated expansion of universe. An acceleration requires energy; This made us think that the accelerated expansion of the universe involves some physical energy. Since no further information than its existence was known about this energy, it was named as dark energy.

The law governing creation of space is a law of creation (section 2.2), which we do not know; The energy involved in this creation is an energy of creation. It is not a physical energy; hence, we do not have to brood over dark energy looking for its physical source.

2.4 Spin

We proposed a picture of spin in the first chapter. However, the unification of this picture with the existing understanding of spin (for example, spin matrices, spin-orbit coupling and so on) is yet an open challenge. I hope it can be met. In the next section 2.4.1, let us analyse a consequence of this picture of spin.

2.4.1 Quantum spin and the classical dynamics

Let's take a hydrogen-like atom's orbital with quantum numbers, $n = 2$, $l = 1$ and $m = 1$ represented by its wave function, ψ_{211}.

$$\psi_{211} = \frac{1}{8\sqrt{\pi}} \left(\frac{Z}{a_0}\right)^{\frac{3}{2}} \rho e^{\frac{-\rho}{2}} sin\theta e^{\pm i(\phi - \omega t)}$$

where Z is the atomic number of the atom, a_0 is the Bohr atomic radius, θ is the polar angle and $\rho = r/a_0$ where r is the radial coordinate.

The two parts of this wave function representing the two parts of the electron are,

$$\psi_1 = \frac{1}{8\sqrt{\pi}} \left(\frac{Z}{a_0}\right)^{\frac{3}{2}} \rho e^{\frac{-\rho}{2}} sin\theta \; cos\left(\phi - \omega t\right) \quad (2.1)$$

$$\psi_2 = i\frac{1}{8\sqrt{\pi}} \left(\frac{Z}{a_0}\right)^{\frac{3}{2}} \rho e^{\frac{-\rho}{2}} sin\theta \; sin\left(\phi - \omega t\right) \quad (2.2)$$

According to hypothesis 2, the squares of these two parts represent the extended existence of the two parts of the electron. The squares are

$$\psi_1^2 = \frac{1}{64\pi} \left(\frac{Z}{a_0}\right)^3 \rho^2 e^{-\rho} sin^2\theta \; cos^2\left(\phi - \omega t\right) \quad (2.3)$$

and

$$\psi_2^2 = -\frac{1}{64\pi} \left(\frac{Z}{a_0}\right)^3 \rho^2 e^{-\rho} sin^2\theta \; sin^2\left(\phi - \omega t\right) \quad (2.4)$$

These are waveforms with two full waves (along ϕ) surrounding the nucleus of the atom. These are the two parts of the electron surrounding the nucleus. The time dependant variation of the two parts is told to be the spin of the electron (section 1.4). The amplitudes of the two wave forms grow and fade sinusoidally and complimentarily such that their sum is conserved at all times, at any given point in space. In other words, at every point in space, the electron transforms from its one component form to the other component form. This is the spin of this electron.

Because of this spin, we observe one more interesting dynamics of the electron. It is the revolution of the two parts of the electron around the nucleus with time, as represented by the equations 2.3 and 2.4. This is the classically expected sense of revolution of the electron around the nucleus. For an electron with zero m value (for example a $1s$ or a $2s$ electron), its two parts are spherically symmetric about the nucleus. Hence, the intensification and rarefaction of these two parts do not result in any such apparent revolution. It is only for those electrons with non-zero m value, do we find an apparent revolution; the direction of revolution depends on the sign

2.4. SPIN

of m value. Now, let's calculate the angular momentum associated with these revolutions.

2.4.2 Calculating the angular momentum

Both the parts of the electron have exactly the same physical significance. Hence, for calculating the physical parameters like angular momentum, the contributions from both the parts have to be considered equivalently. The two parts have opposite signs (eqns. 2.3 and 2.4). Hence, part 2 has to be made positive before adding the two parts together. This is accomplished by multiplying the wave function with its complex conjugate. The classical equation for calculating the angular momentum is

$$L = m_e r^2 \omega \qquad (2.5)$$

where m_e is the mass of the electron, r is the radius of revolution and ω is the angular velocity of the electron. The mass, m_e, of the electron is distributed between its two parts, described by such eqns. as eqns. 2.3 and 2.4. For an electron with non-zero m value, such as the $2p$ electron described by eqns. 2.3 and 2.4, the revolution is about the z-axis, with the azimuthal radius r_a given by

$$r_a = r \sin\theta \qquad (2.6)$$

The angular velocity, ω_ϕ of the distribution is given as

$$\omega_\phi = \nu \phi_\lambda \qquad (2.7)$$

where ν is the frequency of spinning and ϕ_λ is the angle covered by one wavelength of the distribution (squares of the real and imaginary parts of the wave function, such as equations 2.3 and 2.4) along ϕ, given as

$$\phi_\lambda = \pi/m \qquad (2.8)$$

where m is the non-zero magnetic quantum number of the electron. The angular frequency of a wavefunction, ν_ψ is given as $\nu_\psi = E/h$ where h is Planck's constant. Since spinning involves squares of real and imaginary parts of the wave function, the spinning frequency ν is twice ν_ψ.

$$\nu = 2 * E/h = \frac{Z^2 m_e e^4}{4\epsilon_0^2 h^3 n^2} \qquad (2.9)$$

where E and e are the energy and electric charge of the electron, Z is the nuclear charge and n is the principal quantum number of the electronic state, which is 2 in our chosen case. Substituting the above expressions for ν and ϕ_λ in eqn. 2.7 for angular velocity,

$$\omega_\phi = \frac{\pi Z^2 m_e e^4}{4\epsilon_0^2 h^3 n^2 m} \tag{2.10}$$

Let's substitute for r and ω from eqns. 2.6 and 2.10 in equation 2.5 to get the equation for angular momentum as

$$L = \frac{\pi Z^2 m_e^2 e^4 r^2 \sin^2\theta}{4\epsilon_0^2 h^3 n^2 m} \tag{2.11}$$

Since the electron is spread over space in two parts as described by equations, 2.3 and 2.4, we have to integrate eqn. 2.11 over the volume occupied by the electron, to get the angular momentum of the electron.

$$\begin{aligned}L &= \int_{r=0}^{\infty}\int_{\theta=0}^{\pi}\int_{\phi=0}^{2\pi} \psi_{nlm}^* L \psi_{nlm} r^2 \sin\theta \, dr \, d\theta \, d\phi \\ &= \frac{\pi Z^2 m_e^2 e^4}{4\epsilon_0^2 h^3 n^2 m} \int_{r=0}^{\infty}\int_{\theta=0}^{\pi}\int_{\phi=0}^{2\pi} \psi_{nlm}^* \psi_{nlm} r^4 \sin^3\theta \, dr \, d\theta \, d\phi\end{aligned} \tag{2.12}$$

We are considering only the basis eigenfunctions and not their superpositions. Since $\int_0^{2\pi} d\phi = 2\pi$, the above equation can be written as,

$$L = \frac{\pi^2 Z^2 m_e^2 e^4}{2\epsilon_0^2 h^3 n^2 m} \int_{r=0}^{\infty}\int_{\theta=0}^{\pi} \psi_{nlm}^* \psi_{nlm} r^4 \sin^3\theta \, dr \, d\theta \tag{2.13}$$

The numerical evaluation of the above integral for L of a few hydrogen orbitals with non-zero m values are presented in table 2.4.2. The conventional evaluation of angular momentum as $L_l = \sqrt{l(l+1)}\hbar$ and the ratios L/L_l are also presented in the table.

Table 2.1: Calculation of angular momentum

n l m	L	$L_l = \sqrt{l(l+1)}\hbar$	L/L_l
2 1 1	3.160e-34	1.491e-34	2.119
3 1 1	8.420e-34	1.491e-34	5.646
3 2 1	4.212e-34	1.491e-34	1.631
3 2 2	3.159e-34	2.583e-34	1.223

The ratio, L/L_l apparently approaches unity with increasing value of n and decreasing value of $n-l$. However, the deviation from unity has to be pondered upon. Probably, we are missing out something here; something like spin orbit coupling?

Thus, quantum spin causes an apparent revolution of electron distribution in an atom, something like which is expected from the classical point of view. We would like to generalize this concept to any other dynamics and propose the following postulate.

Postulate 4: *All the dynamics, either subatomic or macroscopic, result from the coordinated quantum mechanical spins.*

2.5 Gravity and its effects

Planets go around the sun in orbits, influenced by the gravitational attraction between the sun and the planets. The orbital velocity of a planet around the sun decreases with the increasing distance of its orbit from the Sun. This is as given by Kepler's laws (also by the combination of Newton's laws of gravitation and Newton's laws of motion). If the orbit is nearly circular, the orbital speed is found to be inversely proportional to approximately the square root of the distance from the sun. The orbital velocities of the planets in our solar system are plotted in figure 2.1. Such curves showing the orbital speeds of objects as a function of their distance from the centre of the system are known as rotation curves.

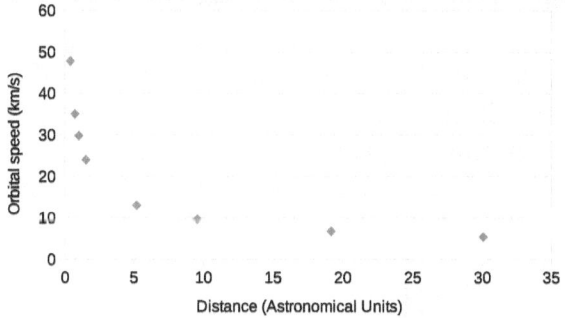

Figure 2.1: The orbital speeds (rotation curve) of planets in our solar system

What about the rotation of stars in galaxies? Stars are also expected to have a similar rotation curve with respect to their distance from the galactic centre. A typical galaxy looks like the galaxy M33, the image of which is shown in the background of figure 2.2. Unlike in the solar system, which has a central gravitating sun with well defined boundaries, a galaxy has a collection of stars distributed over a wide region. However, almost every galaxy has a supermassive black hole at its centre. The rotation curve computed with such distribution of all matter known to exert gravity including inter-stellar gas in the galaxy is shown in figure 2.2, with the label "expected from visible disk".

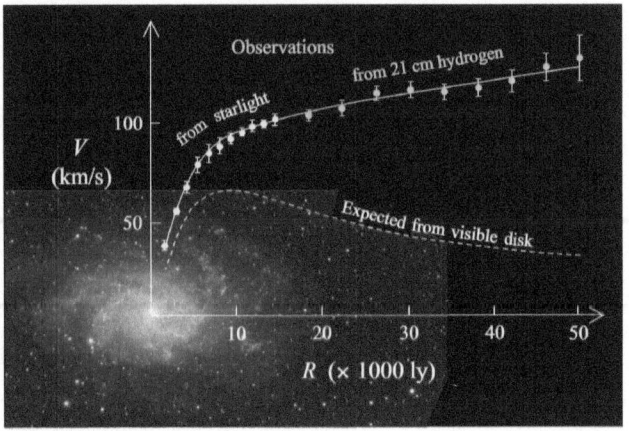

Figure 2.2: Rotation curve of galaxy M33

What is the actual rotation curve observed in the galaxies? The observed rotation curve is very different from the computed curve and is shown in the figure, with the label "Observations". The speed does not decrease as expected, with the distance of stars from the centre of galaxy; in fact, it increases with distance in some galaxies as here. In other galaxies, the speed remains nearly constant, if not increasing with distance as seen in figure 2.3 showing the rotation

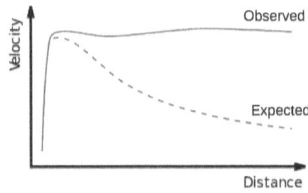

Figure 2.3: The rotation curves of a typical galaxy

2.5. GRAVITY AND ITS EFFECTS

curves of a typical galaxy. Such observed rotation curves can be explained only if gravity is assumed to be exerted by some entity that could not be seen and distributed throughout the galaxy and even well beyond it. We knew that matter exerts gravity. Hence, this unseen entity exerting gravity was assumed to be some matter. Since it is not visible by any means other than the gravity it exerts, it was named as 'dark matter'. It is observed that there is a 'halo of dark matter' spread in and surrounding every galaxy. In other words, every galaxy is embedded in a halo of 'dark matter'.

There have been ground based experiments and space experiments to detect any 'dark matter' particle; however, any such particle has not yet been detected. 'Dark matter' is still considered to be some mysterious matter. Let us see how the concepts discussed in chapter 1 can help in explaining this additional gravity observed in galaxies.

2.5.1 Gravity being explained

The special feature about gravity introduced in chapter 1 is that it is exerted by space also in addition to matter that had been known to exert gravity according to conventional theories of gravity. There is a lot of space in the cosmos. Hence, gravity exerted by space is our candidate to explain the additional gravity observed in the cosmos.

Recap on how space exerts gravity: We saw in section 1.3.3 (postulate 1) that gravity is the force of attraction existing between quanta of subtle source. We also found that a quantum of subtle source exists with every quantum of space and every particle. This is how gravitational force of attraction acts between every quantum of space and every particle. When, because of this attraction, the particles agglomerate to the levels of celestial objects like stars, planets and their satellites like moon, this force becomes measurable and was identified as force of gravity acting between material objects. However, the force of gravity exerted by space is not noticeable, at least at the celestial scales. Let us see whether gravity exerted by space can be measurable at cosmological scales and make it a candidate to explain 'dark matter'.

Space as dark matter

Since the literature available so far use the term 'dark matter', we also use it (within quotes) while referring to existing information on 'dark matter', although we believe that it is actually space. The measured density (mass density) of 'dark matter' is reported[1] to be approximately 7×10^{-25} g/cm^3, in the region of Milky Way galaxy, where our sun is present. This is very less compared to the typical mass densities of normal matter in the form of liquid or solid, which are of the order of a few g/cm^3. Hence, the apparent mass density of space in view of the gravity it exerts is about 10^{25} times smaller than the mass densities of normal matter. Now, let's calculate the mass equivalence of space extended to two different extents in the cosmos, one, our solar system and the other, our galaxy, and compare them with the mass of normal matter in those volumes.

Solar system

The radius of earth's orbit around the sun is 1.496×10^8 km. The mass equivalence of space of spherical shape with this radius, calculated with the above mass density of space, is around 8×10^{12} kg. Although this appears to be huge, it is around 10^{-18} times smaller than the mass of sun, which is approximately 2×10^{30} kg. Hence, the gravity exerted by this space is negligible when compared to the gravity exerted by the sun and hence, it does not produce a measurable change in the orbit of earth around the sun. The gravitational effects of space within the orbits of other planets in the solar system are also too less to cause noticeable change in their orbits. This is the reason why the planets in the solar system are observed to have the orbital speeds computed by applying Newton's laws on normal matter in the solar system (figure 2.1).

Galaxies

The galaxy where our sun is present is called Milky Way. If we see the night sky from a place away from the influence of other lights than those of stars, we will see our galaxy as a band of milky light, since we view the galaxy from its lateral interior. Hence, the name,

---
[1] arXiv:0907.0018v2

2.5. GRAVITY AND ITS EFFECTS

Milky way. The image of Milky way galaxy as seen from its outside is similar to the image of galaxy M33 given in figure 2.2. The radius of this galaxy is around 4.7×10^{17}km. The radius of sun's orbit around the centre of this galaxy is about 2.57×10^{17}km. The mass equivalence of the spherical volume of space with radius equal to this orbital radius of sun, calculated with the density of 'dark matter' near Sun is about 4×10^{40}kg. The density of 'dark matter' has been observed to actually increase as we move towards the centre of any galaxy. Hence, the mass of 'dark matter' in this volume would be larger than that calculated above. This calculated mass is equal to the mass of more than 20 billion suns. The mass of all the normal matter in the whole of galaxy added together is around 90 billion suns. Now, these two masses are comparable unlike the mass equivalence of space being negligible as compared with the mass of normal matter within the solar system. This is the reason why gravity exerted by space in a galaxy causes considerable increase in the orbital speeds of stars. Volume grows as the cube of distance from the galactic centre and similarly will grow the contribution of gravity exerted by space as we move towards and even beyond the periphery of a galaxy. This is what we see in figure 2.2, this gravity causing the orbital speeds to increase with distance from the centre of galaxy. Towards and beyond the periphery, the gravitational contribution by space outreaches that by normal matter. In the overall universe, the mass of normal matter is estimated to contribute around 16% and the mass of 'dark matter' to around 84% of the total mass of the universe[2].

'Dark matter halo'

Every galaxy is observed to be immersed in a halo of 'dark matter'. The density of 'dark matter' is observed to be maximum at the core of the galaxy and decrease as we move outwards.

We found in section 1.3.2 that gravity compresses space. The more a quantum of space is inside a volume of space, the more is the compression it experiences and the smaller it becomes. This implies that the density of subtle source increases in this region. The density of subtle source is seen as the density of 'dark matter'. Thus, if we consider a volume of space, for example, that of the 'dark

---

[2]Information on this can be found in the internet.

matter halo' of a galaxy, the density of the subtle source is highest in the middle and decreases towards the periphery. This explains the observed density profile of the 'dark matter halo'.

In fact, the 'dark matter halo' cannot have a boundary, since space exists everywhere. However, the density of space quanta is lesser in the intergalactic space because of the lower gravity there and this is seen as lower 'dark matter' density there.

2.5.2 Interpreting curvature of spacetime

Understanding world lines

The trajectory of an object is the description of its position as a function of time. The trajectory of an object in a space-time graph is called the world line of that object. Spacetime graph with all the three spatial dimensions and one dimension for time is four dimensional, which is impossible to depict in three dimensional space. Hence, it is customary to limit spatial dimensions to one or two in spacetime diagrams so that spacetime can be illustrated in two or three dimensions respectively. Let us choose to represent one spatial dimension, which is represented by a horizontal axis and time represented by a vertical axis.

Let us assume an object to be stationary. This means that its position is constant with respect to time. This will be illustrated by a vertical world line as seen in plot A of figure 2.4. If the object moves rightward with constant velocity, its world line is a straight line slanting rightward; if it moves leftward with constant velocity, its world line is a straight line slanting leftward. These are illustrated in plots B and C in the figure. If the object accelerates, its worldline becomes non-linear (curves) in the direction of acceleration (Plots D and E in the same figure).

Let us assume two objects, one at rest and the other accelerating toward it. Their world lines are shown in figure 2.5, object A being at rest and object B moving toward object A. This will be the situation if object A is very massive and object B is light in mass and if there is a force of attraction between the two objects.

In fact, the shapes of world lines are relative. The world lines in figure 2.5 are as seen from the reference frame of A. From the reference frame of B, the world line of A curves towards the world line of B as seen in figure 2.6.

2.5. GRAVITY AND ITS EFFECTS

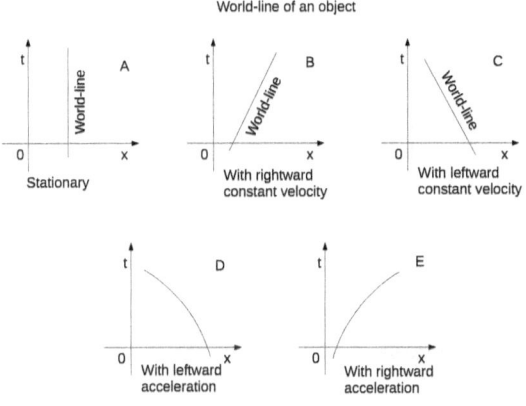

Figure 2.4: World-line examples

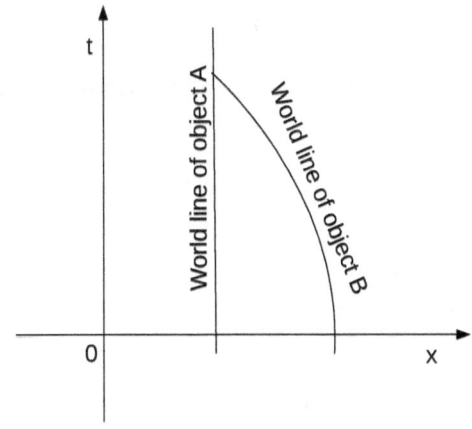

Figure 2.5: An object accelerating towards another

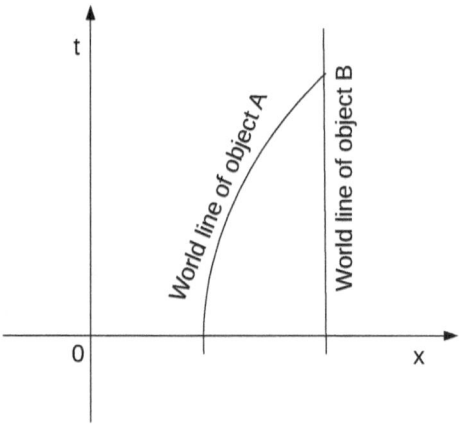

Figure 2.6: Acceleration between two objects as seen from reference frame of B

Thought experiments on gravity

Let us assume four cases, two near earth and two deep in space.

- We simply stand on earth. We feel the gravitational force pressing our feet against the earth. If we drop a ball, it accelerates towards the ground.

- An elevator is accelerating in deep space, in the direction of its top surface. We are standing in the elevator with our feet on the elevator's bottom surface. We feel a force pressing our feet against the bottom surface. This is the same feel that we felt standing on earth. Here again, if we drop a ball, it accelerates towards the bottom surface of the elevator.

- We are in deep space without any acceleration. We don't feel any force acting on us. If we drop a ball, it hovers in space.

- We are in an elevator that is freely falling due to earth's gravity. Inside the elevator, we don't feel the presence of gravity. If we gently drop a ball, it hovers inside the space of elevator.

From the first two conditions, we find that standing on earth's surface is equivalent to accelerating in deep space. From the next two conditions, we find that free fall due to earth's gravity is equivalent to being in an inertial frame in deep space!

According to non-relativistic mechanics, spacetime is flat. This means, in our two-dimensional spacetime diagrams discussed earlier,

2.5 GRAVITY AND ITS EFFECTS

the constant position lines are straight vertical lines as seen in figure 2.7

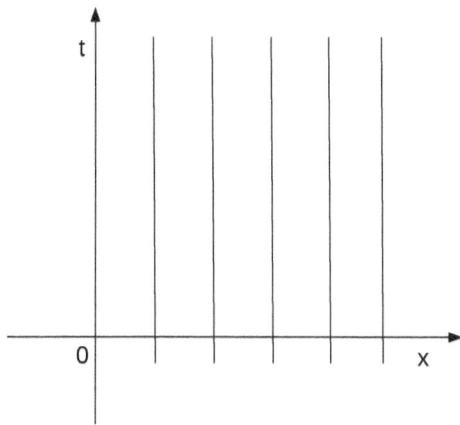

Figure 2.7: Constant position lines of a 'flat' space

Consider figure 2.5, where object B accelerates towards object A because of a force of attraction between them. Let the force of attraction between them be gravitational. This is a special case since the world line of object B is same irrespective of its own mass, which can be understood from equation 2.15. The acceleration due to gravity is $\frac{GM}{r^2}$ in equation 2.15 for the object with mass m (irrespective of the value of m) and $\frac{Gm}{r^2}$ for the other object with mass M (irrespective of the value of M). Now, recall the idea that acceleration due to gravity is equivalent to being at rest in deep space. Then, can we say that the world line of object B in figure 2.5, which is curved due to gravity, represents the object's being at rest? Towards this, Einstein made the bold proposal that gravity curves spacetime. Then, the constant position lines in this curved spacetime would be curved as shown in figure 2.8. He proposed that gravity is not a force and the warping of spacetime caused by gravity makes gravity to appear as a force. However, curvature of spacetime by gravity is not easily conceivable. Can the concepts on gravity introduced in chapter 1 help in this direction?

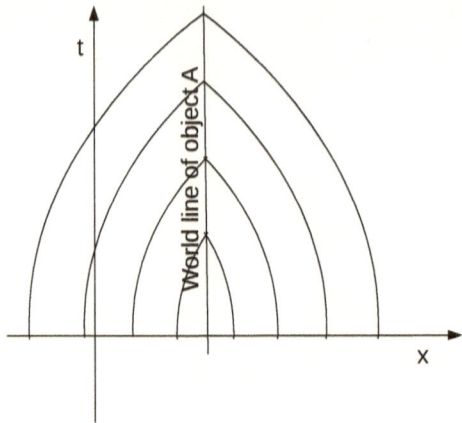

Figure 2.8: The constant position lines curved by the gravity of object A

2.5.3 Corroborating with our philosophy

According to postulate 2, particles are created in a region where the gravitational pressure caused by the surrounding exceeds a limit and this causes a flow of space from the surrounding towards this region. Since space is the medium in which any movement is enacted, an object being stationary in a flowing space is actually moving with the velocity of the flowing space. This is represented by the curved constant position lines, as in figure 2.8 and interpreted as curvature of spacetime.

How far will this rush of space towards core extend? If it is left unattended, it is equivalent to the reversal of creation of space. Recall the 'laws of creation' discussed in section 2.2. According to this law, space should be continuously created at suitable pace and place to feed the conversion of space into particles. In fact, it is not stopped with the compensation of space vanishing at the core; more space is created at an accelerated pace to cause accelerated expansion of space, which was discussed in section 2.3.1.

According to postulate 1, gravity is a force. According to postulate 2, there should be a matter-space density below which gravity causes compression of space and beyond which it causes curvature of spacetime. Compression of space can be viewed as curvature of space. The presence of such a threshold matter-space density has to be verified and the corroboration of the new philosophy on gravity with the very much proven mathematics of relativity has also to be

2.5. GRAVITY AND ITS EFFECTS

verified.

2.5.4 Contraction of space by matter

We found that space gets densified by the surrounding matter and space as observed in cosmological scales (section 2.5.1). Can it be observed at terrestrial scales? We speculated that below the threshold matter-space density, gravity causes curvature of space (section 2.5.3. Then, this curvature should be observable with small objects on the terrain. Probably, this is observed causing refraction of light.

Let us assume space is quantized as a three dimensional array of cubic space quanta. We consider a rectangular prism as shown in figure 2.9. The space quanta are compressed inside this object because of its higher density than the surrounding. This is illustrated in the figure. Suppose a ray of light travels towards this object in the surrounding medium along the diagonal of the space quanta. This is the direction of light with respect to space quanta. The ray will maintain this direction with respect to space quanta inside the object also. Because of maintaining this direction with respect to the space quanta , the ray bends towards the normal of the object's surface while entering the object, as seen in the figure. This is what we observe as refraction. If this is the case, the refractive index of a

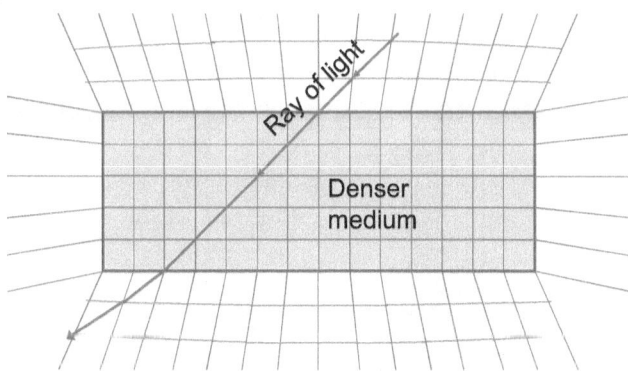

Figure 2.9: Refraction of light by densification of space quanta

medium should be proportional to the mass density of the medium. This is verified in figure 2.10. The deviation in the trend with a few materials may be due to the way in which mass density is distributed within the medium.

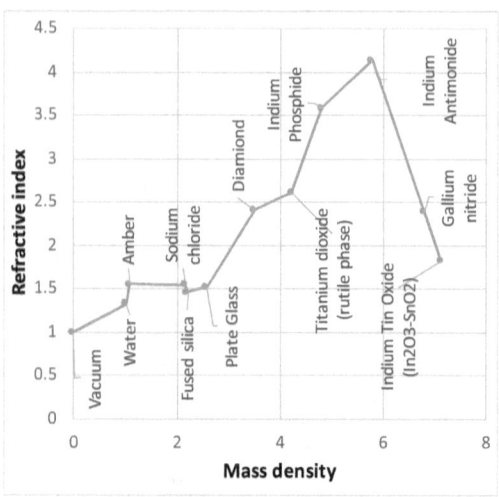

Figure 2.10: Refractive index

The fundamental unit of length is the dimension of the space quanta. Hence, it appears to change from place to place depending on the compression experienced by space quanta. The wavelength of light, which is measured in this unit of length is invariable. Hence, it appears for an observer to reduce in a medium of higher density since the dimension of space quanta there is less.

Similarly, speed is the number of space quanta traversed per unit time. Hence, the speed of light appears for an observer to reduce in a medium of higher mass density. However, for light itself, it is the same velocity in both the media since the number of space quanta traversed by it is the same in both the media per unit time and it is the same direction in both the media since the direction with respect to space quanta is same in both the media.

2.5.5 Gravitational and inertial masses

According to Newton's second law of motion, the force, F acting on an object is given as

$$F = ma \tag{2.14}$$

where a is the acceleration of the object and m is the constant of proportionality. The constant of proportionality m is characteristic of the object. We call this constant as mass of the object. According to this equation, for a given magnitude of force, lower will be the acceleration for higher the mass, which means higher will be the inertia against change in its state of motion (change in velocity). Hence, m in this equation is known as the 'inertial mass' of the object.

According to Newton's law of gravitation, there is a force of attraction, F between any two objects. Newton identified that this force of attraction is responsible for an object to fall on earth and that it acts between the earth and any object near it. He discovered the mathematical equation for the gravitational force between any two objects to be,

$$F = \frac{GMm}{r^2} \tag{2.15}$$

Here, M and m are known as the gravitational masses of the two objects. In this equation, the constant of proportionality is G. Experiments of the best possible accuracy so far show that the inertial mass of an object appearing in equation 2.14 equals its gravitational mass appearing in this equation.

It is a mystery among physicists why the constant of proportionality m (inertial mass) appearing in equation 2.14 should reappear as a factor in equation 2.15 for gravitation. Science always encourages questioning and grows in the process. For generality, physicists expect, in eqn 2.15, a parameter different from the inertial mass.

Gravity and inertia - a speculation

Suppose that we try to move an object. There should be an inertia against this movement caused by the densification of space quanta inside the object, since new space quanta have to be densified in the front end of the object and some space quanta have to be rarefied from the rear end. This picture is analogous to the Higg's mechanism of objects gaining mass due to the drag caused by 'Higg's field'.

2.6 What is more?

There may be more consequences to the philosophies discussed so far. We can continue exploring them in coming times. So far we have discussed physical sciences. Our physical sciences deal with physical entities. Even what we call as life sciences deal with biophysics and biochemistry. Hence, they also actually form part of physical sciences. However, there are subtler entities like mind; even our life itself is a subtle factor. We have to look for other sciences to understand those faculties. We will see them in the next chapter.

Chapter 3

The Science Beyond

There are three types of evidences or sources (pramaanaas) of our knowledge. They are

1. **Kaakshi (Direct):** What we observe directly through our senses, mind etc. For example, we see objects through our eyes, feel them by touch, hear sounds and so on.

2. **Anumaanam (inferred):** What we infer from what we have already known. For example, we inferred gravitational force by observing falling objects. We could not see the subatomic world directly. However, we have inferred a lot about them by observations like spectral emission from atoms etc.

3. **Aahamam (Scriptural):** The knowledge through scripture.

In fact, the second (inferred) type of knowledge can be based on first (direct) and/or third (scriptural) type of knowledge, known respectively as kaakshi anumaanam and aahama anumaanam.

The above classification is, in fact, from scripture.

3.1 The purpose of this chapter

The knowledge dealt in the previous two chapters are of type one and part of type two (inferred from direct). This chapter discusses the third type, the Scriptural knowledge and also sometimes type two knowledge, inferred from scripture.

3.1.1 Limits of physical sciences

Let us accept a postulate that a thing can understand or create things that are grosser (less subtle) than itself. This means, for example, man can create a computer. However, a computer cannot create another computer like itself or better or a human cannot create another human. Computer creating another computer can occur only in fictions; such events are fictitious, not real.

Similarly, our intelligence developed physical sciences, which is the study of the physical entities that are grosser (less subtle) than intelligence itself. We cannot explain mind, intelligence and other things that are subtler than the physical entities through physical sciences. Hence, we have to go beyond physical sciences to know the subtler things; they are the scripture that we have to seek.

We have evolved a lot in physical sciences and in technologies. They provide a lot of comfort to us, which in fact is not eternal. We actually seek from our inner depth something eternal; eternity and eternal bliss. This is not provided by any physical means. We have to go through spiritual path to get the state of all eternity. We will have a comprehensive look on spiritual path in section 3.3.2. Before that, let us introduce the scripture. While discussing the scripture, we will also remark some similarities between scripture and some aspects discussed in the first two chapters.

These are some reasons for us to know scripture and the purpose of this chapter.

3.2 The scripture

Scripture in India are known as Vedha. Vedha talks about four things; dharma, artha, kaama and moksha. Dharma means virtues, virtues in our thought, speech and action. Artha describes the physical life with dharma. Kaama mentions the pleasures achievable by following dharma and artha. Moksha describes getting relieved from physical bondages and reaching the all eternal divine. Tamil and Sanskrit are the oldest languages told to be given by the God, Lord Shiva directly. Vedha are available in both these languages. In Tamil, dharma, artha, kaama and moksha are known as, aram, porul, inbam and veedu respectively. Vedha consists of two parts. The first part is known as karma kaanda, which means 'the part describing duties'. It deals in detail about dharma, artha and kaama (aram, porul and inbam) and indicates to some extent about moksha (veedu). Although it is a part of vedha, it is commonly referred to by the generic name vedha.

The second part of vedha is known as Gnaana kaanda, which means 'the part describing knowledge or wisdom'. It includes dharma, artha and kaama and deals in detail about moksha. In the process, it also describes what the world is, how and why the worlds are created, who created and so on. It consists of upanishads. Upanishads are called as vedhantham, which means end or conclusion of vedha (Vedha + antham, antham means end).

The tamil version of karma kaanda is Thirukkural. The tamil version of vedhantha is Saiva Sithandham. 'Saiva' is the adjective form of 'Siva', meaning, related to Lord Siva. The meaning of Sithandham will be discussed in the next section.

Arul means divine grace, compassion, mercy and the likes. Arulaalar means the person with arul. God and arulaalaas (plural form of arulaalar) are proponents of vedha. Arulaalar Umapathy Sivachariyar describes saiva sithandham as the clear version of vedhantham.

There is a lot to write about scripture. However, we, in this part, limit to a concise description of them. In this chapter, we will get a brief description of Saiva sithandham. I bow at the feet of all the arulaalaas and gods; May them bless with correct presentations of scripture here.

3.3 Saiva Sithandham

Saiva Sithandham is given by Lord Shiva through Kayilaya Paramparaa[1]. Lord Shiva appears in twenty five main moorthams (forms). Lord Shiva in his Guru form (Bestower of wisdom) is known as Lord Dhakshinamoorthy (figure 3.1). The four rishis (saints) around his feet are Arulaalaas Sanaha, Sanarkumara, Sathya Gnaana Dharshini and Paranjothi Munivar; Lord Dhakshinamoorthy transfers Gnaana to them. Paranjothi Munivar transferred the Gnaana to his disciple Shri Suvedha vana Perumal at his age of one and a half and named him Meikandaar, which is the Tamil translation of Sathya Gnaana

---
[1]Kayilaya parampara means Lineage of Kayilaayam or Kailash. The Kailash in the Himalayas is a representation on earth of the divine Kayilayam.

Figure 3.1: Lord Dhakshinamoorthy

Dharshini (Guru of Paranjothi Munivar). Arulaalar Meikandaar is the proponent of Saiva sithandham in Tamil. He presented the work called 'Siva Gnaana Bhodham', which consist of twelve sutras. Sutras are concise statements (aphorisms). His disciples elaborated on the sutras adding thirteen more works on Saiva sithandham. Thus, there are, in total, fourteen prime works on Saiva sithandham. Let us have a glimpse of Saiva sithandham.

Saiva sithandham declares the factual of three things and declares them as eternal and without origin. The three things are The God (Pathi), the souls (Pasu, also called as aanma or aathma, meaning us) and darkness or ignorance (paasam). Any one of these did not create any of the other two. As apparent, the God and darkness are opposite in nature to each other. The God and the souls are similar in nature. However, a soul is atomistically small whereas the God is indescribably great, glorious, omnipresent, omnipotent, omniscient and so on. Scripture describe God's greatness by his eight qualities. They are, (i) being the self, unperturbed, (ii) pure bodied, (iii) naturally conscious, (iv) completely conscious, (v) naturally away from darknesses, (vi) greatly graceful, (vii) endlessly powerful and (viii) boundlessly blissful. The God and the souls have three attributes. They are, wish (ichchaa), wisdom (Gnaana) and deeds (Kriya). The darkness does not have any of these attributes. It is inanimate.

For the soul, the two other things present are, the God and the darkness. A soul being an atomistic part of the god, its ichcha (wish) towards darkness is unfulfilled. Hence, it enters into darkness to satisfy this ichcha. Although the soul enters the darkness, the God maintains his link with the trapped souls. God's ichcha is to restore the souls from darkness. In the trapped condition, the Gnaana (wisdom) and Kriya (deed) of the souls become inactive. God's gnaana and Kriya are towards restoring the souls towards him. Towards this, he creates three levels of bodies (thanu or sareera) to the souls, instruments (karanaas) in these bodies, worlds (bhuvanaas) and experiences (bhohaas) through these, for the souls. These are created by God from his power (Shakthi) called as Maya (pronounced as Maayaa). The Maya is with three levels of subtleties. The subtlest Maya is called as Pure Maya (Shudhdha Maya). The Maya with intermediate subtlety is called as Vidhya Maya. The grossest one is called as Nature Maya (Prakruthi Maya).

God's Kriya are told to be fivefold. They are, (i) creation (of thanu, karana, bhuvana and bhoha), (ii) sustenance of the creations, (iii) withdrawal of creations, (iv) keeping his presence with the souls and in everything in the creations not explicitly visible to the souls and finally (v) showing himself to the souls at appropriate times and allowing them to merge with him.

3.3.1 God's creations from Maya

The purpose of God's creation of thanu, karana, bhuvana and bhohaa from his sakthi 'Maya' is to get the souls relieved from darkness and reach him. Including darkness, there are three hurdles in the path towards God. The other two are the deeds (karma) done by the souls and the Maya itself since the soul has ichchaa towards creations from Maya also. In fact, the soul's nature is to adhere to something or other always. Originally, it adhered to darkness, then adheres in parts to darkness, Maya and God and finally it completely merges at the feet of God. The soul's path is, getting relieved gradually from darkness with increasing merger at the feet of God.

God creates from the prakruthi maya, subtle faculties including mind, intelligence, ego and sitham and the physical entities like matter, force and so on. He creates from the Vidhya Maya, time, laws and subtler faculties like inner ego, inner intelligence and inner mind. The laws include physical laws, which physical sciences understood and are trying to understand and subtler laws. An example of subtler laws is the laws of karma (deeds)[2]. Inner ego can be thought of as freewill given to the souls to take an autonomous decision.

Of the three bodies given to the souls, the innermost is known as causal body (kaarana sareeram) and is constituted by time, laws, inner ego, inner intelligence and inner mind. God resides in the inner

---

[2]Karma are divided into three parts; sanchitha karma, praraptha karma and aahaamiya karma. Sanchitha karma is the collection of all the deeds a soul has done in all its previous births. God chooses a part of sanchitha karma, which he finds suitable for the souls to experience the consequences of, in its next incarnation; this portion of sanchitha karma is known as praraptha karma and is bestowed with a birth. The consequence of bad deeds (called as paabha) is suffering as decided by the god; the consequence of good deeds (punniya) are of two types; worldly enjoyments or getting closure to the god; it depends on the attitude of the souls and as decided by the God. While enjoying the consequences of praraptha karma in a birth, the soul also does new deeds, which are called as aahaamiya karma and these add to the sanchitha karma. Why God bestows the consequences of karma to the kartha (doer, the soul) is for the soul to get the clarity, to get purified and get relieved from the bad attributes of the three hurdles in the spiritual path and progress towards god. The first karma executed by a soul is its entering into the darkness; It is called as moola (original or root) karma.

3.3. SAIVA SITHANDHAM

mind of every soul. The next level of body is known as the subtle body (Sootchuma sareeram) and is constituted by mind, intelligence, ego and the five senses. The third level of body (grossest) is the physical body (sthoola sareeram). Everyone of us know the existence of our physical and subtle bodies; with the progress in spiritual path, we will start knowing the existence of the causal body.

Sithandhas and Saiva sithandham As mentioned above, mind, intelligence, ego and siththam are some subtle creations from prakruthi maya. Siththam is the subtlest of these. Mind receives information from our sensory organs; intelligence processes the information; ego arises to do something based on this process[3]. Knowledge acquired based on this process is stored in sitham; Sitham contains the collection of all the knowledge acquired based on the past experiences. Sithandham is the conclusion arrived at, based on the knowledge in the sitham (Andham means end or conclusion). Thus, sitham+andham means the conclusion arrived at the sitham. Everyone would have one's own sithandham based on one's own experiences. The experiences could include reading and teachings.

Saiva sithandham is the conclusion of scripture. Arulalar Umapathy Sivachariyar describes saiva sithandham as the ended (settled or concluded) conclusion. There is no further development required from this.

Overlap with physical sciences

We found above according to scripture that the physical entities are created from prakruthi maya; It can be noted that this is equivalent to the concepts introduced in the earlier chapters that physical entities like space and particles are created from 'the subtle source'. We have also found in scripture that time is created from Vidhya Maya, which is subtler than the prakruthi maya. Similar is the idea, which we formulated in earlier chapters, that time is created from a source that is subtler than the source of other physical entities (hypothesis 7). These are some similarities we can remark, which show overlap between scripture and physical sciences.

---

[3] Ego is as such not a bad entity as is conventionally thought of. Ego is necessary for our development. It is required to do good things so that we progress in our spiritual path. It should not be used to do bad deeds.

Some of my associates commented that readers may get the impression that I formulated the concepts in the first two chapters with a deliberation to cause an overlap between the scripture and the physical sciences. Hence, for clarity on this aspect, I would like to give a rough timeline and the flow of logics that led to the postulates and hypotheses given in the first two chapters. When I was an undergraduate in the second part of 1980's, I inferred something about planetary motions. It is that, if we suppose that the sun and the planets are going in circular orbits about a central point, the motion of planets as observed from the sun's frame of reference would be precessing elliptical orbits, which is observed. I also supposed therefore that electrons and the nucleus of an atom also move in circular orbits about a central point, analogous to the solar system. In the mid 90's when I was introduced to quantum mechanics, I was pondering on the imaginary part of the wave functions; As a result, I concluded that the two parts of a wave function should represent two mutually opposite parts of the particle, which that wave function represents. Later, when I heard that when a sufficiently energetic gamma ray photon comes close to the nucleus of an atom, the photon gets converted into an electron and a positron, because of the very high field strengths near the nucleus. I interpreted this in a different way. Since every particle in an atom (the particles in the nucleus and the electrons) consist of two mutually opposite parts, the central part around which they revolve may contain something neutral, 'neutral' with respect to both of these two mutually opposite parts; this neutral stuff should be the 'agent' responsible for the conversion of a photon into material particles. Later, I learnt Saiva sithandham. This made me interpret this neutral part to be a subtle thing. Then came the ideas of space being created from the subtle source, gravity exerted by subtle source, conversion of space into particles and so on. This is a rough timeline and flow of ideas that caused the development of the philosophies introduced in this book.

The science beyond The physical sciences deal with the physical entities and to some extent with time. The scripture deal with God, souls, darkness and all the creations including the physical entities and the subtle ones like mind. Hence, they are The Science Beyond. They also present the spiritual path, which is essential to go beyond the limits of the physical world.

3.3.2 Spiritual path

Spiritual path is the one through which the soul gets rid of the three obstacles (beginning of section 3.3.1) and reunites with the Lord. The path consists of four stages. They are called as Sariya, Kriya, Yoga and Gnaana. The first stage 'sariyaa' is related to physical deeds like doing good deeds, physical deeds towards worshiping the god and so on. The second stage 'Kriya' involves mind, for example, involving our mind in the worship. The third stage 'Yoga' involves attaining merger with the god by crossing some obstacles through some steps including meditation. The fourth stage 'gnaana' is the destiny of reaching the God crossing all the obstacles.

In fact, each of the four stages contains the components of all the four stages. Thus, in total, there are sixteen components in the spiritual path as shown in table 3.2. The first stage consists of all

Gnaana in Sariyaa	Gnaana in Kriya	Gnaana in Yoga	Gnaana in Gnaana
Yoga in Sariyaa	Yoga in Kriya	Yoga in Yoga	Yoga in Gnaana
Kriya in Sariyaa	Kriya in Kriya	Kriya in Yoga	Kriya in Gnaana
Sariyaa in Sariyaa	Sariyaa in Kriya	Sariyaa in Yoga	Sariyaa in Gnaana

Figure 3.2: Spiritual stages

the components containing Sariya, the 'L' shaped cells in the table with darkest background colour. The second stage, Kriya, has the components in the next inner level 'L' cells with intermediate grey colour background. The third stage, Yoga, is constituted by the components in the third inner 'L' cells with the lightest grey colour background. The final stage, Gnaana, is the Gnaana in Gnaana stage shown in the cell with white background.

The third column 'Yoga' is described in Sage Pathanjali's 'Yoga Suthras', which describe Yoga. Yoga consists of eight components listed below.

1. Yama: Five things to practice, which indicate five things not to do

(a) Ahimsa: non-violence

(b) Satya: truth (non-lying)

(c) Astheya: non-stealing

(d) Brahmacharya: non-indulgence in sensual pleasures

(e) Aparigraha: non-greed

2. Niyama: Five things to practice

    (a) Soucha: purity

    (b) Santhosha: contentment

    (c) Tapas: discipline

    (d) Svadhyaya: exploring the self

    (e) Eeshvara Pranidhana: surrender to God

3. Aasana: Postures conducive for spiritual practices as well as physical and mental well-being

4. Pranayama: Breathing practices promoting general health and spiritual practices

5. Prathyahara: Withdrawal of the mind from worldly matters and senses; in essence, it means getting relieved from obsessions with worldly matters;

6. Dhaarana: enlightening concentration on suitable things, which causes progress in spiritual path

7. Dhyaana: meditation or continued dharana

8. Samaadhi: getting in some sense equal to the origin; sama means equal and aadhi means origin

Aasana that is the third component of Yoga, which are physical postures that promote spiritual practices (and health too) as mentioned above is commonly referred to as Yoga; However, it is to be noted that Yoga comprises of all the eight components listed above.

Now, let us discuss the sixteen components of spiritual path.

1. Sariya in Sariya: Doing physical activities towards devotion, like visiting the temples, cleaning the temples and serving the devotees

2. Kriya in sariya: Participating in the worship of God

3. Yoga in Sariya: Feeling a merger with the God worshipped

4. Gnaana in Sariya: Attaining some wisdom through these sariya deeds

5. Sariya in Kriya: Physical deeds like collecting flowers and other materials required for the worship of god

6. Kriya in kriya: Worshipping the God internally, in our mind

7. Yoga in kriya: Feeling a merger with the God worshipped internally

8. Gnaana in kriya: Attaining some wisdom through these kriyas

9. Sariya in Yoga: Yama, niyama, aasana and pranayama, which are the first four components of yoga, listed above

10. Kriya in yoga: Prathyahara and dharana, which are the fifth and sixth components of yoga.

11. Yoga in Yoga: Dhyana that is the seventh component of yoga

12. Gnaana in yoga: Samaadhi that is the eighth component of yoga

13. Sariya in Gnaana: Listening to the scripture from a sacred Guru

14. Kriya in Gnaana: Pondering on the scripture

15. Yoga in Gnaana: Getting a clarity on scripture

16. Gnaana in Gnaana: perceiving all that is learnt in scripture by direct experience; complete union at the feet of the Lord and enjoying the five activities (mentioned before the beginning of section 3.3.1) of the Lord.

A guru initiates [4] the saadhahaas [5] in the spiritual path in three main stages. The first stage puts the person in the sariya stage. The second initiation makes the soul go through the Kriya and Yoga stages. The final initiation leads the soul to the Gnaana stage. For

---
[4]The process of initiation is known as providing dhiksha.
[5]Practitioners in the spiritual path are known as saadhahaas.

a practical account of the experiences by spiritually evolved souls, one can read the books like,

- 'The Autobiography of a Yogi' by Shri Paramahamsa Yogananda.
- 'Apprenticed to a Himalayan Master: A Yogi's autobiography', by Sri M.

The Yogoda Satsanga Society of India - Self-Realization Fellowship established by Shri Paramahansa Yogananda has gurus to train saadhahaas in Kriya Yoga. Sri M also initiates suitable persons into the spiritual path and conducts many *satsangs*[6].

3.3.3 Can it be a probabilistic evolution?

Scripture says that the world is created by the God. According to some evolutionists, the evolution of inanimate world and the animate entities like plants and animals is probabilistic. Could the *world* have evolved to its current state just by random probability? Let us examine this by considering the cases of camouflaging insects; one, that of a grasshopper having a wing-like structure with striking similarity to that of a neem-leaf, the other is another insect with its structure having a striking similarity with that of a dried and curled mango leaf (see figures 3.3 and 3.4).

Figure 3.3: Grasshopper with neem-leaf structure

How could an insect have the look of a leaf in its body? Everyone understands that the purpose of this is to camouflage with

---

[6]*Sat* means those that are eternal; sang means union or mingling. Thus, *Satsang* means the union or mingling for the eternal.

Figure 3.4: An insect looking like a dry mango leaf

the leaves of the trees and thereby to protect it from its predators. Did the insect see its predators and the leaves and wish to develop this appearance in its body? Even if it wishes, can it develop this structure by itself? At least, humans can have such wishes; Even for them, it is not possible to evolve their physique according to such wishes.

The plants, the insects and the predators may have their own lines of evolution, if at all. How can a feature unique to one line of evolution appear in the evolution of another line? Does this arrangement not necessitate a common being who would have created these three groups, viz, the trees with their leaves, the insects having the structure of these leaves and the predators of these insects? If one would argue that there could have been a common code to create the leaf structure in the trees and in these insects, it can be refuted by the fact that the leaf structure in the plants are functional for processes like photosynthesis, while those structures in the insects are dumb, only a mimic. Thus, these indicate that the world could not have evolved by random probability.

3.3.4 Withdrawal of creations?

In previous chapters, we found that physical entities are created from subtle sources. In section 2.3, we also wondered whether there would be any event of withdrawal of a created entity, that is, whether any creation is taken back into its subtle source. We have not observed any such withdrawal so far. Now, let us see what scripture say about withdrawal of creation.

Souls go through innumerous cycles of birth and death in their course of (spiritual) evolution. They are told to get tired in this process. Hence, God chooses to give rest to the souls by withdrawing all the creations that are grosser than time and all the souls, into *time*. This event is known as a Maha Pralayam. Taking rest is meaningful only with respect to time. Hence, time is required for the souls' resting to be meaningful. Hence, all the creations are withdrawn into 'time'. Thus, 'time' is always ticking. The time gap between a creation and such a withdrawal is supposedly billions of years (I am not sure about the exact numbers). Hence, 'scientists' may have to wait for such a long period to 'watch' such a mega event! However, we can keep investigating to see whether any such withdrawals happen anywhere locally.

Scripture is vast. I have concisely presented it in this chapter according to the limited knowledge I have. Let us know scripture as required and evolve in spiritual process.

Chapter 4

Overview

We found new insights into physical sciences in the first two chapters. The important one is the introduction of the entity called 'subtle source'. This has an equivalent entity in the scripture, discussed in chapter 3, which is called as 'Maya'. According to me, scripture 'Saiva sithandham' is the philosophy that can be referred to verify the validity of a new concept, if a direct or indirect description of it can be found in it.

The first two chapters are open for evaluation by experts in corresponding fields of physical sciences. The third chapter is for anyone interested to know scripture. Readers may pay attention to all the chapters or exclusively to either the first two chapters or the third chapter, according to their likings.

There are a few aspects in the new physical philosophies introduced, which need further clarity. For example, gravitational force is told to cause flow of space due to creation of particles and also told to act as a force on matter; Will the effect of gravity on matter be the cumulative of these two? The mathematics of these philosophies are to be developed. Let us hope that we can do it in the future.

www.ingramcontent.com/pod-product-compliance
Lightning Source LLC
Chambersburg PA
CBHW030526220526
45463CB00007B/2738